Here are 50 more black inventors for you to discover. Research - who, what, why, when and maybe even how! Happy Exploring!

JAMES S ADAMS

ALEXANDER P ASHBOURNE

LEONARD C BAILEY

JAMES A BAUER

ANDREW J BEARD

GEORGE E BECKETT

ALFRED BENJAMIN

HENRIETTA MAHIM BRADBERRY

PHIL BROOKS

HENRY BLAIR

CHARLES B BROOKS

OSCAR E BROWN

JOHN A BURR

BURRIDGE & MARSHMAN

R A BUTLER

DEWEY SANDERSON

CLATONIA J DORTICUS

SOLOMON HARPER

ROBERT FLEMMING JR

DAVID A FISHER

BENJAMIN F JACKSON

_____
_____
_____
_____
_____
_____
_____
_____
_____
_____
_____
_____
_____
_____
_____
_____
_____
_____
_____
_____
_____
_____

JOHN A JOHNSON

GLENVILLE T WOODS

ADOLPH SHAMMS

W H RICHARDSON

JOESEPH C PRICE

REV. HENRY MCNEAL TURNER

BENJAMIN J GREGORY

JAMES E WEST

MARK DEAN

ST ELMO BRADY

DR CHARLES RICHARD DREW

MARGARET KNIGHT

BESSIE BLOUNT GRIFFIN

THOMAS ELKINS

FREDRICK JONES

DANIEL HALE WILLIAMS

_____

_____

_____

_____

_____

_____

_____

_____

_____

_____

_____

_____

_____

_____

_____

_____

_____

_____

_____

_____

_____

_____

THOMAS L JENNINGS

PERCY JULIAN

MARC HANNAH

DR CARLTON GYLES

DR JUDE IGWEMEZIE

WILLIAM PEYTON HUBBARD

MARY JONES DE LEON

JUDY W REED

JANET EMERSON BASHEN

ARCHIE ALPHONSO ALEXANDER

VALERIE THOMAS

ELLEN EGLIN

JOSEPH DICKINSON

Well done! - 50 more inventors that you know.
Which was your favorite?

_____

_____

_____

_____

_____

_____

_____

_____

_____

_____

_____

_____

_____

_____

_____

_____

_____

_____

_____

_____

_____